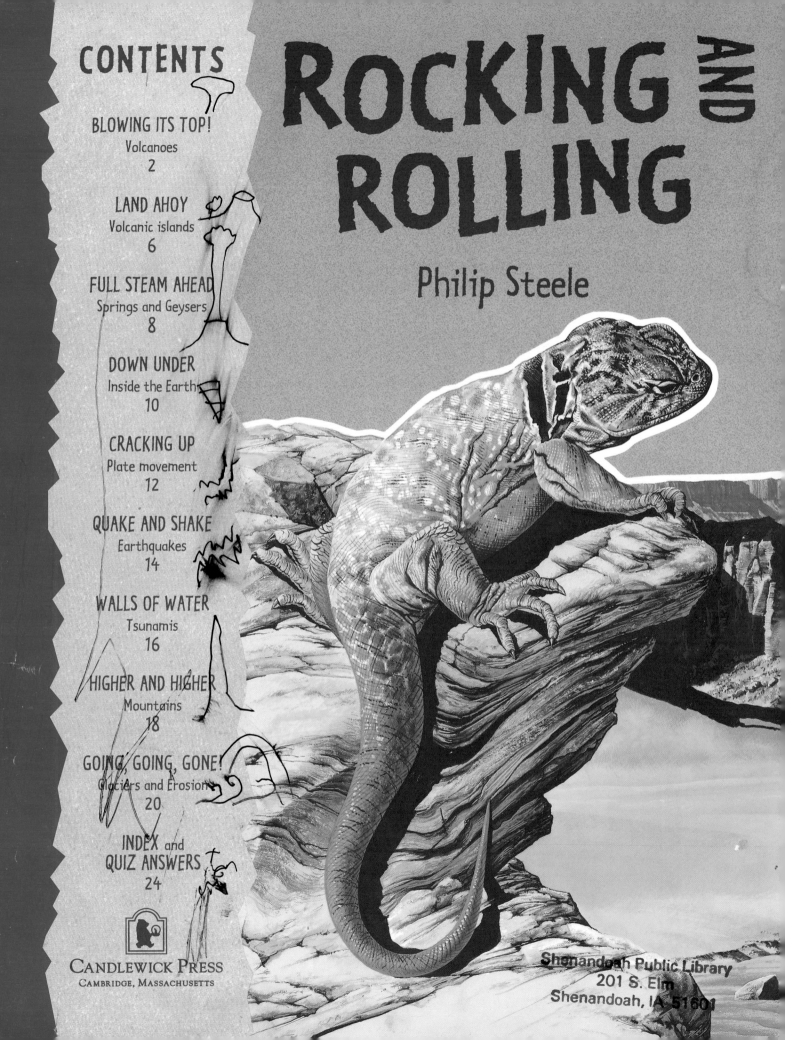

CONTENTS

ROCKING AND ROLLING

Philip Steele

CANDLEWICK PRESS
CAMBRIDGE, MASSACHUSETTS

1 There are volcanoes all over the world. Scientists who study them are called volcanologists, and they try to figure out when volcanoes are going to erupt.

2 Sometimes the scientists climb down into the mouth of a volcano to see what's going on.

3 When they do this, they have to wear special clothing to protect them from the heat and the poisonous gases that a volcano gives off.

4 They use a machine called a seismometer to measure the ground shaking, and they use a special thermometer to calculate the temperature inside the volcano. This information helps the volcanologists tell how close the volcano is to erupting.

5 When it gets too dangerous for people to go inside the volcano, a robot is sent down instead. It can do many of the same jobs as the scientists, and it doesn't have to wear any special clothing!

BLOWING ITS TOP!

1 This is a very strange-looking mountain. What's happened to the top of it? There's just a big hole where its peak should be.

2 The hole is called a crater, and it tells us that the mountain is a volcano.

3 And now this volcano is erupting! Clouds of smoke and gray ash are billowing out, filling the air with stinking poisonous gases.

4 Boiling rivers of red-hot rock are gushing out of the crater and streaming down the sides of the volcano. The rock is called lava, and it comes from deep under the ground.

5 Far beneath the volcano there's a great hollow. It's called a chamber, and it's a place where gases and hot gooey rock called magma collect.

4 VOLCANOES

6 When a volcano erupts, it's because the gases and magma are forcing their way up through giant cracks in the ground and out through the crater.

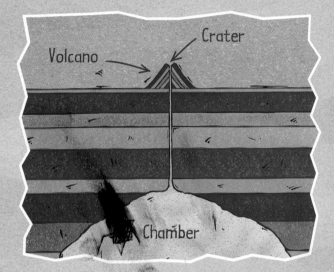

Volcano

Crater

Chamber

7 And when magma reaches the earth's surface, we call it lava.

8 Eventually the volcano stops erupting. The lava cools and hardens into black rock, and the mountain is quiet — until the next time!

LAND AHOY

1 Sparkling blue seawater laps gently at the shores of a beautiful island. Long sandy beaches stretch as far as the eye can see...

2 but wait a minute. Why is the sand black? And look at the crater on top of that mountain.

3 Well, the color of the sand tells us what it's made of—black volcanic lava. And that crater shows that the island is really just the tip of a huge volcano.

6 **VOLCANIC ISLANDS**

4 The island began forming long, long ago when magma burst up through a crack in the seabed.

5 As the hot gooey magma met the cold seawater, it hardened into shiny black rock.

6 Eruption followed eruption and the rock slowly grew into a mountain. Then one day, its top broke through the surface of the sea—our island had been born!

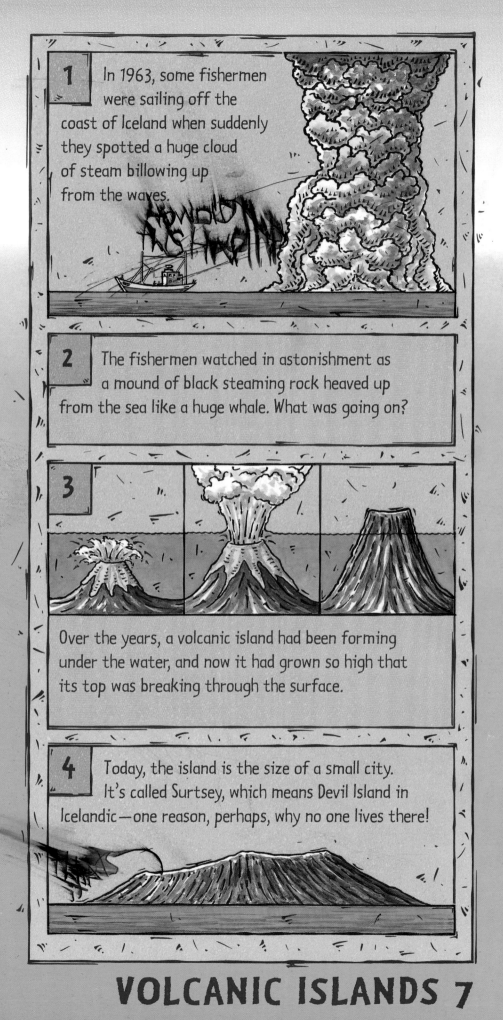

1 In 1963, some fishermen were sailing off the coast of Iceland when suddenly they spotted a huge cloud of steam billowing up from the waves.

2 The fishermen watched in astonishment as a mound of black steaming rock heaved up from the sea like a huge whale. What was going on?

3 Over the years, a volcanic island had been forming under the water, and now it had grown so high that its top was breaking through the surface.

4 Today, the island is the size of a small city. It's called Surtsey, which means Devil Island in Icelandic—one reason, perhaps, why no one lives there!

VOLCANIC ISLANDS 7

1 Hot springs are pools or streams of naturally hot water. They're found where rain-water has collected underground and been heated by hot rocks deep inside the earth.

2 People all over the world make use of underground hot water. In Iceland, the water is piped into people's homes to provide heating.

3 In Budapest, Hungary, the water from hot springs is fed into open-air swimming pools.

4 But people aren't the only ones who like warm water. There are hot springs high in the snowy mountains of Japan, where monkeys called macaques go swimming!

1 Whoooosh! A great fountainlike jet of hot water and steam spurts up out of the ground. This is a geyser.

2 The water and steam are coming from the underground pools that form when rain seeps down through cracks in the rocks.

8 SPRINGS

FULL STEAM AHEAD

3 The deeper you go into the earth, the hotter it gets. And when an underground pool gets hot enough, the water boils and turns into steam.

4 The steam then forces its way back through the rocks, pushing any water in the cracks up in front of it—and blasting out into the air as a geyser.

5 The world's tallest geyser is in Wyoming, in Yellowstone National Park. It's called the Steamboat, and it can shoot to the amazing height of 380 feet!

DOWN UNDER

1 Our planet Earth is huge—about 3,960 miles from the surface to the center. Walking this far would take you about 55 days and nights.

2 You wouldn't be able to walk to the center of the earth, though, as it's incredibly hot. It's at least 9,000°F, which is the same temperature as the surface of the sun.

HO HO
HO HO

Outer core

Inner core

3 Earth has four layers. The top one is called the crust and it's made of rock. It's about 25 miles thick under the land, but only about 5 miles thick beneath the ocean.

4 The mantle is next. It's also made of rock, but it's so hot that parts have melted into magma and are as gooey as oatmeal.

Crust

← Mantle

5 Beneath the mantle is Earth's core. This is made of metal and has two layers—an outer and an inner core.

6 The outer core is runny because it's so hot. But although the inner core is even hotter, it's solid. Why? Because the other three layers are pressing down on it and the weight is enough to squash it solid!

1 People have always dreamed of digging down to the earth's center. But so far no one has invented a machine that would survive the heat.

2 The world's deepest mineshaft is in Carletonville, South Africa. But even at 2.25 miles deep, it's little more than a pinprick in the earth's crust.

3 The deepest a person has ever been is the Mariana Trench, a crack in the Pacific Ocean floor. In 1960, a two-person sub dived down into the trench—nearly 7 miles below the sea's surface.

4 In 1970, Russian scientists began drilling a hole into the earth's crust. By 1992, they'd dug more than 7.5 miles down. The temperature at the bottom was more than 400°F—yet the hole was still nowhere near the mantle layer!

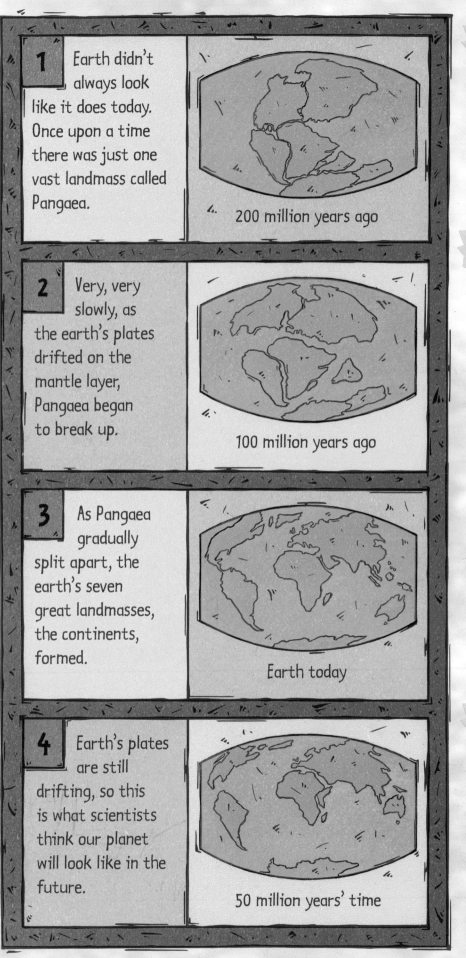

1 Earth didn't always look like it does today. Once upon a time there was just one vast landmass called Pangaea.

200 million years ago

2 Very, very slowly, as the earth's plates drifted on the mantle layer, Pangaea began to break up.

100 million years ago

3 As Pangaea gradually split apart, the earth's seven great landmasses, the continents, formed.

Earth today

4 Earth's plates are still drifting, so this is what scientists think our planet will look like in the future.

50 million years' time

1 If all the oceans disappeared, Earth would look just like a jigsaw puzzle made up of lots of big pieces.

2 The pieces are called plates, and there are about 20 of them. They float on the earth's mantle, moving very, very slowly—between 1 and 8 inches per year.

3 Sometimes the plates move apart and gooey magma rises up from the mantle to fill the gap. The magma cools and hardens to form new land or ocean floor.

12 PLATE MOVEMENT

CRACKING UP

4 Sometimes two plates push against each other. The edge of one plate may slide under the other and melt back to magma...

5 or the two plate edges may crumple up into a range of mountains— even below the ocean.

PLATE MOVEMENT 13

QUAKE AND SHAKE

1 Just a few minutes ago this truck was speeding along the road. Then, suddenly, there was a terrifying roar, and the ground opened up—an earthquake!

2 The most serious earthquakes happen deep underground, along the edges of the earth's plates.

3 Usually, the plates stay jammed close together. But from time to time a plate breaks away.

4 This makes the ground shudder and shake. Sometimes it can even split wide open.

14 EARTHQUAKES

5 These shudders can be felt thousands of miles away because they spread out from the earthquake's center like the ripples from a stone thrown into a pond.

6 Every year, there are 40,000 to 50,000 earthquakes that are strong enough to be felt. However, only about 40 of them are big enough to cause any damage.

1 It's fairly easy for scientists to tell where most earthquakes will take place, but it's hard to figure out when. So people in high-risk areas such as Japan and the west coast of the United States must always be prepared.

2 Architects try to design buildings that will stand up to the ground shaking. For example, the Trans-America building in San Francisco is a slender pyramid because this is such a sturdy shape.

3 In Japan there's a National Disaster Prevention Day each year, when everyone practices what to do during an earthquake. Volunteers spend the day learning how to rescue people from fallen buildings.

4 And in California, children have earthquake drills at school. One thing they're taught is to take shelter under their desk—it's the safest place to be if the roof falls in.

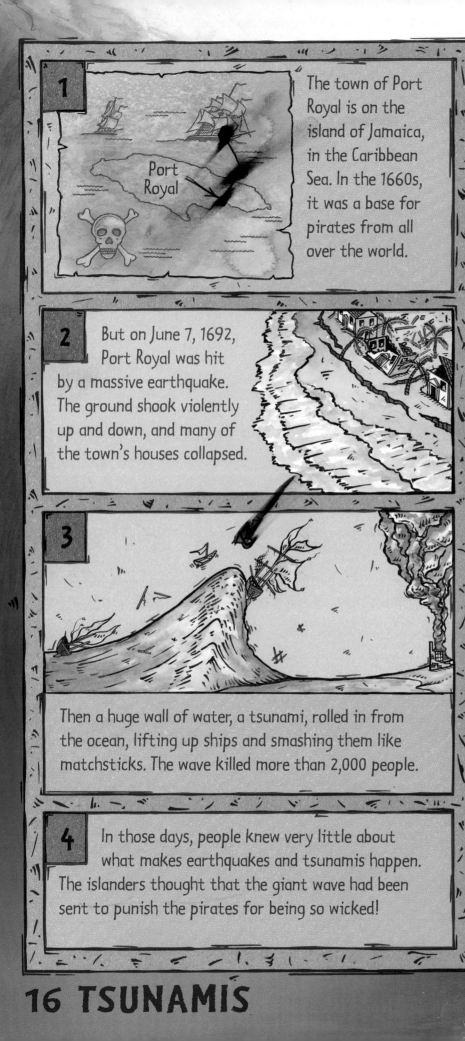

1 The town of Port Royal is on the island of Jamaica, in the Caribbean Sea. In the 1660s, it was a base for pirates from all over the world.

2 But on June 7, 1692, Port Royal was hit by a massive earthquake. The ground shook violently up and down, and many of the town's houses collapsed.

3 Then a huge wall of water, a tsunami, rolled in from the ocean, lifting up ships and smashing them like matchsticks. The wave killed more than 2,000 people.

4 In those days, people knew very little about what makes earthquakes and tsunamis happen. The islanders thought that the giant wave had been sent to punish the pirates for being so wicked!

1 Most waves are made by the wind blowing over the sea. This is no ordinary wave, though. It's called a tsunami, and it was started by an earthquake.

16 TSUNAMIS

WALLS OF WATER

2 A big earthquake is a lot like a huge bomb going off. The force of the explosion can create a tsunami that travels thousands of miles through the ocean.

3 When the tsunami is in deep ocean water, its top may be only 8–12 inches above the surface.

4 But as it rolls on into shallower water near the coast, the tsunami is forced upward into a gigantic wall of water—sometimes it can be even higher than an apartment building!

HIGHER AND HIGHER

1 Welcome to the top of the world. The Himalayas are the highest mountain ranges on Earth, and the tallest Himalayan peak is Mount Everest.

2 Everest already soars to a height of 29,029 feet. But next year it will be a tiny bit higher—it's still growing.

1 Long ago, people made up stories to explain where mountains came from. In one tale they were elephants that had been magically turned into stone. But, of course, mountains are really made by the rocking and rolling movements of the earth's crust.

2 There are three main kinds of mountains. Fold mountains, like the Himalayas, form when the Earth's plates crunch into one another, and layers of the crust are pushed up into loops and bumps.

3 The Himalayas started forming around 53 million years ago when the earth's plate carrying the land that is now India began crunching upward into the rest of Asia.

4 Inch by inch, India pushed northward. And over tens of millions of years, the plate edges crumpled into the huge ridges, peaks, and valleys we see today.

Himalayas
Mount Everest
CHINA
INDIA

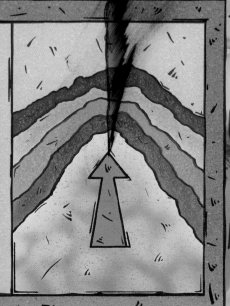

3 Block mountains are made when part of the crust is forced up between two cracks in a plate. These cracks are called faults.

4 Dome mountains happen when magma bulges up beneath the crust. This forces the crust up into a large rounded hump— much like the back of an elephant!

1 High in the highest mountains, it's so cold that snow doesn't melt, even in summer. As more snow falls, it packs down hard into hollows and the snow turns to ice.

2 When a hollow is full, the ice slides down the mountain as a glacier. Most glaciers move very slowly—the fastest creep less than 3 feet a day.

CRUNCH

Glacier

3 Pieces of loose rock are frozen into the bottom of the glacier, and as it edges along, they scrape at the ground underneath. Over the ages, this gritty ice grinds the land away, leaving a deep valley.

4 A valley made by a glacier is a very different shape from one made by a river. This valley was worn away by a river. The fast-flowing water cut through the land making a shape like a V.

rshhh

5 This valley was carved out by a glacier. A glacier is much slower than a river, and like a great bulldozer, the ice cuts out a U-shape from the rock.

GLACIERS 20

GOING, GOING, GONE!

1 It took hundreds of thousands of years to make this enormous valley. But although it happened so long ago, we know what did the work.

2 A mighty river once flowed here. And like all rivers, it carried grit and pebbles that rubbed against the land, slowly wearing it away and carving out a valley.

3 The river has long since dried up and disappeared. But look what it left behind.

4 How come these strange pillars of rock are here? Why didn't the river wear them away?

EROSION 21

5 Well, some rocks are harder than others, and hard rocks break down more slowly than soft ones. The river dried up before it had time to wear these pillars away.

6 The wearing away of the land is called erosion, and it's still going on today. But water isn't doing the work now—so what is?

22 EROSION

7 The answer is blowing in the wind. Day after day it whistles through the valley, picking up grit and sand, and blasting everything it touches.

8 It works like sandpaper, slowly wearing the rocks down and grinding them into weird and wonderful shapes.

9 Rock is much harder than wind and water—yet given time, wind and water are powerful enough to shape the land we live on.

EROSION 23

INDEX

Main illustrations by Chris Forsey (cover, 3, 4–5, 8–9); Christian Hook (16–17);
Ian Jackson (20–23); Mike Lister (14–15); Darren Pattenden (18–19); Luis Rey (12–13);
Ian Thompson (10–11); Peter Visscher (6–7)
Inset and picture-strip illustrations by Ian Thompson
Designed by Jonathan Hair and Matthew Lilly; edited by Sarah Hudson
Consultant: Keith Lye

Text copyright © 1997 by Philip Steele
Illustrations copyright © 1997 by Walker Books Ltd.

First U.S. edition 1997

Library of Congress Cataloging-in-Publication Data is available.
Library of Congress Catalog Card Number 97-539

ISBN 0-7636-0303-1

2 4 6 8 10 9 7 5 3 1

Printed in Italy.

This book was typeset in Kosmik.

Candlewick Press
2067 Massachusetts Avenue
Cambridge, Massachusetts 02140

QUIZ ANSWERS

Page 2 — TRUE
Mars and Venus both have volcanoes.
The biggest is more than 16 miles high.
It's on Mars and it's called Olympus Mons.

Page 6 — FALSE
The volcanic island of Pulau Batu Hairan, off
the coast of Malaysia, appeared in 1988.

Page 9 — FALSE
Using information about when a geyser has
erupted in the past, scientists can figure
out when it's likely to blow its top again.

Page 10 — TRUE
Earth's core and the moon are both about
2,200 miles wide, measured straight
across the middle.

Page 13 — FALSE
They are moving apart at a rate of
about 1 inch a year.

Page 14 — FALSE
Usually, earthquakes last between
a few seconds and a minute.

Page 17 — TRUE
Tsunamis can travel at more than 550 miles
per hour—as fast as a jet plane.

Page 19 — TRUE
The Hawaiian volcano Mauna Kea rises
33,480 feet from the seabed, but only the
top 13,796 feet are above the water. Mount
Everest is 29,029 feet above sea level.

Page 20 — FALSE
Some valleys are formed by movements
of the earth's crust.